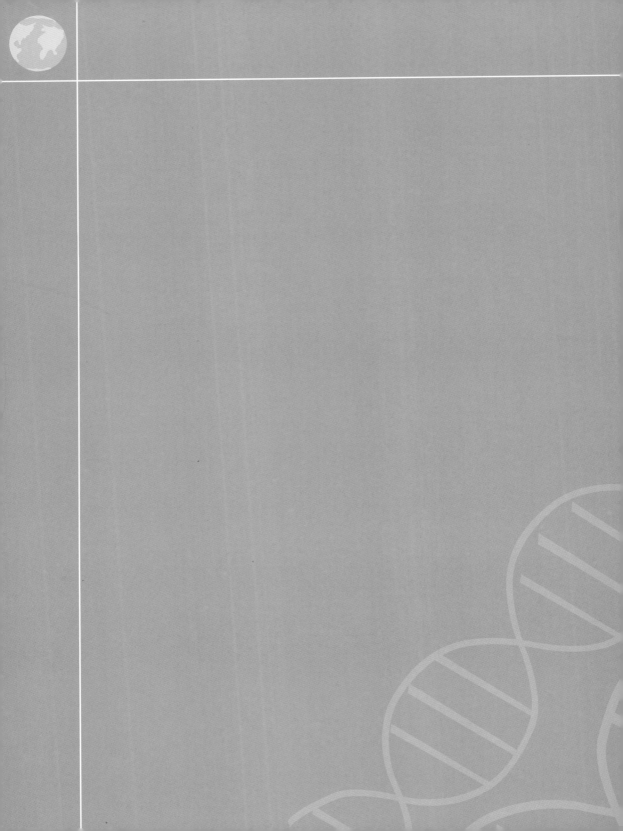

Water Scientists

by William B. Rice

Science Contributor
Sally Ride Science
Science Consultants
Nancy McKeown, Planetary Geologist

Developed with contributions from Sally Ride Science™

Sally Ride Science™ is an innovative content company dedicated to fueling young people's interests in science.

Our publications and programs provide opportunities for students and teachers to explore the captivating world of science—from astrobiology to zoology.

We bring science to life and show young people that science is creative, collaborative, fascinating, and fun.

To learn more, visit www.SallyRideScience.com

First hardcover edition published in 2010 by
Compass Point Books
151 Good Counsel Drive
P.O. Box 669
Mankato, MN 56002-0669

Editor: Mari Bolte
Designer: Bobbie Nuytten
Editorial Contributor: J.M. Bedell
Media Researcher: Svetlana Zhurkin
Production Specialist: Jane Klenk

Copyright © 2010 Teacher Created Materials Publishing
Teacher Created Materials is a copyright owner of the content contained in this title.
Published in cooperation with Teacher Created Materials and Sally Ride Science.
Printed in the United States of America in North Mankato, Minnesota.
092009
005618CGS10

 This book was manufactured with paper containing at least 10 percent post-consumer waste.

Library of Congress Cataloging-in-Publication Data
Rice, William B. (William Benjamin), 1961–
  Water scientists / by William B. Rice.
     p. cm. — (Mission science)
  Includes index.
  ISBN 978-0-7565-4307-5 (library binding)
  1. Hydrology—Study and teaching (Elementary) 2. Water—Study and
teaching (Elementary) 3. Hydrologists—Biography. I. Title. II. Series.
  GB662.3R52 2010
  551.48092′2—dc22   2009035403

Visit Compass Point Books, a Capstone imprint, on the Internet at *www.compasspointbooks.com*
or e-mail your request to *custserv@compasspointbooks.com*

Table of Contents

The Wonders of Water 6
Mohammed Karaji 8
Bernard Palissy 12
Edmond Halley 18
George Hadley 22
Henry Darcy 23
Pelageya Polubarinova-Kochina 27
Environmental Engineer: Alexandria Boehm 29
      Water Scientists at a Glance 30
      Glossary 34
      Water Science Through Time 35
      Additional Resources 38
      Index 39
      About the Author 40

The Wonders of Water

From space, Earth looks like a floating blue ball. It's blue because almost 75 percent of its surface is covered with water. That huge amount of liquid makes our planet unique. Other planets have some water, but no known planet has that much.

Scientists who study water are asking the question, how did it all get here in the first place? For a time, some thought it was carried to Earth on comets. A large number of comets hit Earth more than 4 billion years ago, during what is now called the heavy bombardment period.

Did You Know?

The atmosphere of Earth is a closed system. Matter is rarely added to it or subtracted from it. Molecules of water that existed millions of years ago are the same ones floating around today.

The comet theory had one big problem. Comet water has elements in larger amounts than Earth water. If comets brought the water, then what happened to the extra elements? They can't just disappear.

Another theory is that Earth's water rode in on meteors. So scientists crunched the numbers. They figured out that all of Earth's water could have come from one moon-sized meteor or a bunch of smaller ones.

Many scientists see problems with the meteorite theory, too. So while they continue their debate, other scientists are studying the water itself. From ancient times to today, men and women have been fascinated with water. Here are some of the scientists who led the way and helped us understand the wonders of water.

What Is Water?

Water is chemically bonded hydrogen and oxygen molecules. Each molecule is made from two hydrogen atoms bonded to one oxygen atom. We write water as the equation $H_2+O=H_2O$. These tiny H_2O molecules group together in the billions to make water. Water can exist in three states—as a solid (ice), a liquid (water), or a gas (water vapor).

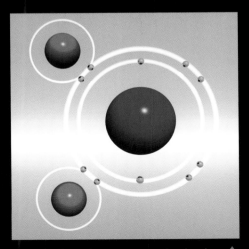

To make a water molecule, the hydrogen and oxygen atoms bond

Mohammed Karaji (c. 953–c. 1029)

In the 10th century, the Middle East was an important center for learning. Men traveled from every corner of the known world to study there. They studied art and music, philosophy and religion, medicine and mathematics. One great man who lived during this time was Mohammed Karaji.

Karaji was born in the city of Karaj, in modern-day Iran. He studied engineering and mathematics. Those who knew him called him Al-Hasib ("the calculator.") While still a young man, Karaji moved to Baghdad, a city in modern-day Iraq. He worked as a mathematician and wrote several books on algebra and geometry. His work was little known until the 19th century when his papers were translated into German.

Later in life, Karaji left Baghdad and moved to an area he described as the mountain countries. Away

⬇ Karaji studied at such learning centers as the Library of Hulwan in Baghdad.

from Baghdad, his focus changed from mathematics to hydrology—the study of Earth's water and how it flows—and hydraulics—the study of fluids.

In the desert region where Karaji lived, finding water was very important. He studied the colors in rock formations, and the growth patterns of plants, to see if he could predict where to find it. Eventually he discovered patterns that consistently led him to the best places to find fresh groundwater.

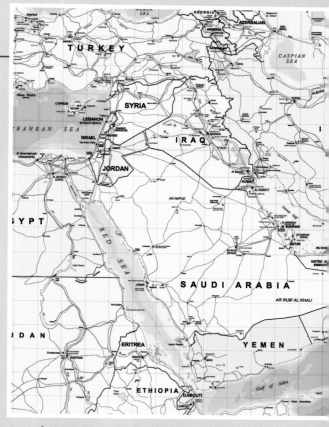

⬆ Baghdad is in modern-day Iraq.

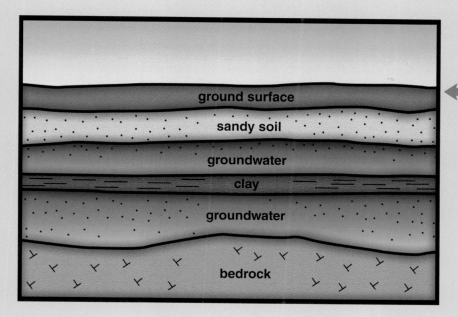

⬅ Layers of soil and rock through which water flows are known as aquifers.

Karaji showed that water underground flowed through aquifers found within layers of rock, sand, or gravel. He also described how fresh water flowed through these underground aquifers and fed wells and springs.

His work allowed farms, villages, and towns to find water much closer to where they were located.

Later he developed a system of tunnels that moved water from a source to where it was

Qanaats

In Arabic an underground irrigation canal is called a *qanaat*. These canals are built when a farm or village needs water. The water comes from a well or aquifer and is moved through underground tunnels. Holes open into the tunnels to allow air to get in and sand and dirt to be removed. From above, these holes look like strings of giant donuts lying on the ground. About 75 percent of the water in Iran flows through qanaats. The country maintains more than 100,000 miles (161,000 kilometers) of tunnels.

The earliest qanaats date back

needed. He proved that water could be moved from place to place, through these tunnels, and used to water crops or bring fresh drinking water to the people.

In the last years of his life, Karaji invented several devices that helped to survey the land and place the tunnels. He also taught farmers how to dig wells, clean dirty water, and keep their irrigation systems working.

Karaji wrote his ideas down in a book titled *Inbat al-miyah al-khafiya*, or *Book on the Extraction of Hidden Water*. This book is the oldest text ever written on the science of groundwater.

Land surveying is a profession that goes back through all recorded history. The same rules of math and science that were used long ago are still used today.

Europeans learned of Karaji's techniques through such trading centers as Enisala in Romania.

Bernard Palissy (c. 1510–c. 1589)

Long before any other scientist, Bernard Palissy discovered that rain was the source of water for rivers and springs. He was the first to describe the water cycle.

Palissy was born in France and spent his youth learning to blow and paint glass. As an apprentice, he was taught to paint portraits, survey land, and make pottery. Like many young men of the 16th century, Palissy worked as a traveling craftsman. Along the way, he acquired many new skills. When he finally settled down, sometime around 1539, he earned his living by using every skill he had learned.

With a growing family to provide for, Palissy turned to pottery as a source of income. He spent 16 years perfecting his craft. His best-known style is called rustic ware—baked clay that is fired at low temperatures and has a

Bernard Palissy was interested in Chinese porcelain and tried to replicate it without success.

very coarse texture. He decorated his pottery with images of reptiles and insects.

Because of the colorful lead glazes he used, the reliefs on each piece appeared to be alive. Each image was detailed, showing the structure of an insect's wing or the perfect tint of a flower's petal. Palissy's distinct style has been imitated by potters throughout the centuries.

Palissy lived at a time when everyone was supposed to be a member of the Roman Catholic Church. Those who weren't Catholic were

⬆ Palissy put animals and plants on his works.

The Water Cycle

Water is always moving from place to place and changing from one state into another. Water that evaporates from the oceans and land floats up into the atmosphere. The evaporated water forms clouds, which move around and eventually drop the water as rain or snow. When the water lands on the ground, it begins to move. It flows into rivers and streams or seeps into the ground. Eventually it all makes its way back into the ocean. This is called the water cycle.

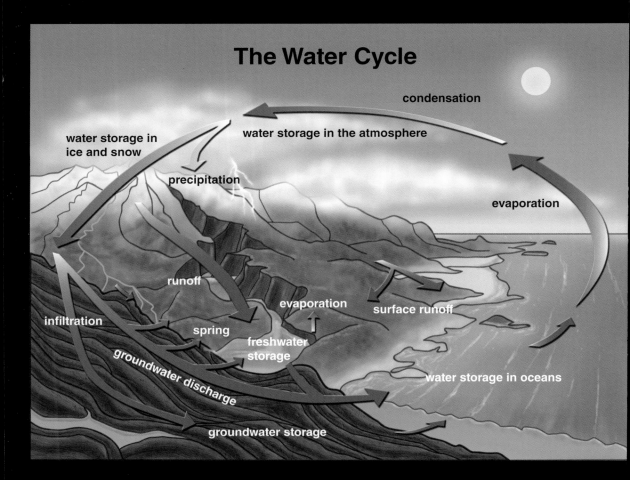

called heathens—people who don't believe in God. Heathens were often sent to prison or put to death. Palissy believed in God but not in the Catholic Church. He decided to become a Huguenot—a follower of John Calvin.

Palissy helped start a new church in his community. Before it had a chance to grow, many of its members were brutally attacked and murdered. Palissy's workshops were destroyed, but Palissy himself was protected by a noble who had taken a liking to Palissy's work.

In 1563 Palissy was allowed to restart his work, this time in Paris near the Louvre. His work caught the attention of the French court, including the queen mother Catherine de Medici. In 1565 he was named "Inventor of rustic pottery to the king and queen mother." He eventually built a rustic pottery grotto—an artificial, cavelike building traditionally set in a garden—for the queen mother's palace.

Over the next two decades, Palissy studied agriculture, geology, paleontology, and other Earth sciences. He gave lectures on natural history and eventually published his lectures.

Palissy's book was called *Discours admirables*, or *Admirable Discourses*. In his book, he argued that rain fell on the mountains and entered cracks in the ground. It flowed downward until something blocked its path and forced it to the surface.

Who Were the Huguenots?

The Huguenots were the French branch of the Protestant reformation—a movement by people who protested Catholic practices and sought to reform the church. They were often persecuted for their faith.

From there, some of the water bubbled up as springs and some made its way into streams and rivers. Palissy also knew that water could sit above sea level, but only if the source of that water was even higher up.

Another of Palissy's theories related to fossils and floods. During Palissy's time, many people believed that the flood mentioned in the Bible had spread various fossils around the world, from the lowest valley to the tallest mountain. They did not believe that the fossils were the remains of dead animals because the animals did not exist in their time. They did not believe that high mountains had once been deep underwater.

Palissy rejected the biblical idea and instead insisted that the fossils were made from animals that had once lived, and that Earth could have been much different millions of years ago. In his lectures, he explained that every fossil lay where the animal once lived.

Continuing Palissy's Work

It seems that no one listened to Palissy's theory. Through the century following his death, scientists continued to argue about the source of the water that flowed into rivers and springs.

René Descartes (1596–1650) thought it came from the ocean. He said that ocean water flowed into channels and was stored

Athanasius Kircherin (1601–1680) thought that the tides pumped seawater into channels. It flowed through these channels and bubbled to the surface as springs.

Robert Plot (1640–1696) thought that air pressure forced water up inside the mountains, where it then broke through the surface and flowed back down.

Covered by layers of soil, the dead animals slowly turned to stone. They lay undiscovered until revealed by erosion, weather, or some other means.

In 1588 Palissy was imprisoned for being a Huguenot. King Henry III offered him his freedom if he converted to Catholicism, but Palissy refused and died in the Bastille a year later.

Inside the Bastille

The Bastille was built in Paris between 1370 and 1383 during the Hundred Years' War. It initially served as a fortress to protect the city from rival armies. After the war, it was first used as a treasury and later became a prison. King Louis XIII (1601–1643) was the first ruler to use the Bastille to hold prisoners.

Its eight towers usually held upper-class criminals—people who had committed treason, political prisoners, dishonest businessmen, and those who were found with forbidden writings, including religious prisoners like Palissy. Later more common criminals were held.

The Bastille was stormed on July 14, 1789, by an angry mob

French Revolution. That day has since been celebrated in France as Bastille Day.

Did You Know?
The word bastille means "castle" or "stronghold" in French.

Edmond Halley (1656-1742)

Edmond Halley may be best known for the comet that bears his name. But he also did important work with water. Halley was the son of a wealthy London soap maker. His father's business was successful because of the increasing popularity of using soap to bathe. The elder Halley also owned a number of properties.

When Halley was 10, the Great London Fire swept through the city, and his father lost almost everything. Luckily he could still afford to send his son to a good school. Halley finished his early education at St. Paul's School, where he excelled in mathematics and astronomy.

Halley continued his studies at Queens College in Oxford. Before he finished his degree, he left England to study the stars of the Southern Hemisphere. He traveled to Saint Helena, a British island off the southwestern coast of Africa. While there, he cataloged 341 stars and a star cluster, and gained

a reputation as a leading astronomer.

In 1678 King Charles II declared Halley a graduate of Queens College. That same year, at the age of 22, he became one of the youngest fellows admitted into the Royal Society of London.

While Halley was studying the southern stars, he was also thinking about weather. He published a paper that charted the path of winds around Earth. The paper also described Earth's winds' effect on weather patterns

Halley's Comet

Edmond Halley studied comets in school. He figured out that several of the comets that were mentioned in written texts as old as 240 B.C. were actually the same one. He used what information was available and figured out when the comet would return. He was right.

By using Halley's mathematical formula, each return of Halley's Comet can be predicted. It passes Earth every 75 to 76 years and can be seen without a telescope. The last time it appeared was in 1986. The comet is expected to return sometime in mid-2061.

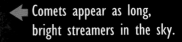
Comets appear as long, bright streamers in the sky.

and monsoons—winds that come every year, last for months, and usually bring a lot of rain.

In that same paper, Halley wrote that heat from the sun affects how weather systems travel around Earth. He also linked air pressure and altitude to changes in local area weather. Because of his work on winds and tides, he is called the founder of geophysics.

Halley was interested in what was beneath the surface of water. He sketched plans for and built a diving bell capable of staying underwater for almost two hours. He and

The Distance From Earth to the Sun

Edmond Halley figured out how to measure the distance from Earth to the sun. Sadly, the measurements could only be taken two times every 100 years. Halley died before the next opportunity arrived. Luckily, he left detailed information for other scientists. When the time came to take the measurements, scientists around the world used Halley's formula. And it worked! Their measurement of 95 million miles (153 million kilometers) is very close to the measurement we have today.

five friends dived 60 feet (18 meters) down in the River Thames. Halley's bell also had a window, so he could see under the water. Later he improved on the design of his bell. The new one could stay underwater for more than four hours.

Halley was fascinated with the work of Bernard Palissy. He agreed with Palissy that rain was the source of water for rivers and springs. But one big question remained. Did enough water evaporate from the ocean, and come down as rain, to account for it all?

Halley measured the amount of water that evaporated from pans on a hot summer day. He assumed that rate was close to what would evaporate from the Mediterranean Sea. Then he calculated what amount would evaporate from the entire sea and came up with 5,280 million tons (4,790 million metric tons) a day.

With that number in hand, he then figured out how much water flowed from

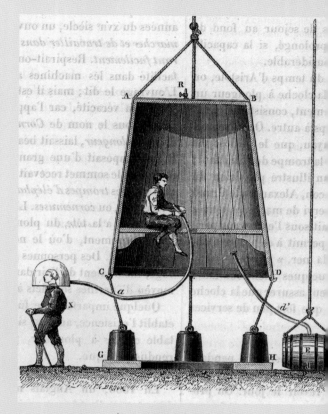

▲ Halley's diving bell

nine major rivers into the Mediterranean. That total was 1,827 million tons (1,657 metric tons), slightly more than a third of the amount lost by evaporation. There was plenty of water to feed the rivers and streams. The other two-thirds of the evaporated water, Halley said, either fell back as rain over the sea or was taken up by plants.

George Hadley (1685-1768)

George Hadley was a British lawyer who much preferred studying physics. He expanded on Edmond Halley's work and explained why trade winds occur. He said that heat from the sun evaporates a lot of water near the equator. That water vapor rises up into the atmosphere and flows north and south. It travels a long distance, and when it cools, it falls back to Earth. As it falls, it pushes air down, creating wind. The rotation of Earth causes the wind to blow from east to west. These east-to-west winds are called trade winds.

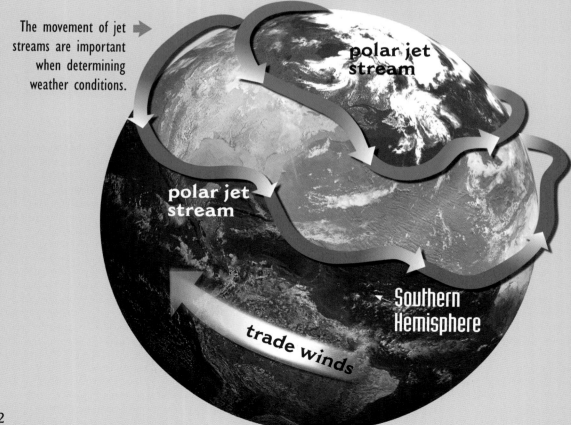

The movement of jet streams are important when determining weather conditions.

Henry Darcy (1803-1858)

Henry Darcy was born in Dijon, France. Later he would play an important role in his hometown's life. Darcy's father died when he was 14, and his mother had to borrow money to keep him in school. He studied science and engineering and graduated at the top of his class.

After graduation he worked for the French government building bridges and roads. His talent and skills soon earned him the position of lead engineer.

At the time, people had to haul water from rivers and streams to use in their homes. Darcy wanted to change all that. He designed a water transfer system that supplied clean water to the city of Dijon.

Did You Know?

Henry Darcy worked on railroad and canal projects as well as bridges and roads.

The system took water from the Rosoir Spring 7.5 miles (12 km) away and moved it through aqueducts into two huge reservoirs. Using only the force of gravity, the water traveled from the reservoirs through 17.5 miles (28 km) of water lines. It ended up in one large public fountain, several major buildings, and 142 public street hydrants. Dijon was one of the first cities in Europe to have such a system.

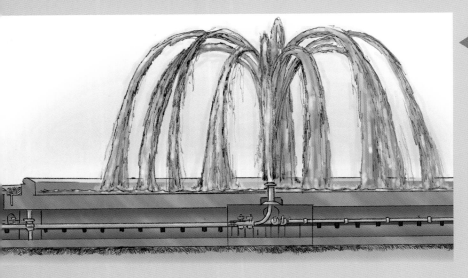

⬅ Because the reservoir was at a higher altitude than the city, Darcy built fountains that used gravity to shoot water into the air.

The Henry Darcy Medal

The Henry Darcy Medal is given to a European water scientist whose outstanding work in water research or water engineering and management honors the legacy of Henry Darcy. The 2009 medal went to Demetris Koutsoyiannis, a professor at the University of Athens and editor of *Hydrological Sciences Journal*, for his work on the effects of climate change on the movements of water.

After his water transfer system was finished, Darcy went on to study other properties of water. His research led to a mathematical equation that defines how much flow and friction is lost when water goes through pipes. His equation is still used today and is called the Darcy-Weisbach equation.

When his health began to fail, Darcy traveled from Paris back to Dijon. While there he experimented with and then published Darcy's law. It defines the rate that water flows through porous materials such as sand. Darcy's law has been the basis for work in groundwater hydrology, soil physics, and petroleum engineering. Darcy's law was Darcy's last work. While on a trip to Paris in 1858, he died unexpectedly of pneumonia at age 54.

Did You Know?

Upon the completion of his reservoir project, Darcy was offered 55,000 francs as payment. He refused and accepted only a gold medal and free water for the rest of his life.

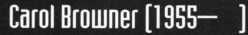

Carol Browner (1955—)

Carol Browner fights for the environment. She works to keep Earth safe for future generations. Her goal is "to leave the world a slightly better place."

Browner grew up near the Florida Everglades. Her parents taught her the importance of caring for our land, water, and air. After college and law school, she went to Washington, D.C., to work with environmental groups.

In 1993 President Bill Clinton made her head of the Environmental Protection Agency. She stayed in that job for eight years, the longest in EPA history.

A big part of her work at the EPA was to protect and clean our water. During her time in office, she made sure that millions more people had clean water to drink. In 1996 she led the fight for the renewal of the Safe Drinking Water Act.

After leaving the EPA, Browner worked in private business. But her heart was still with the environment. She served as chairperson for the Audubon Society and on the boards of the Alliance for Climate Protection and the League of Conservation Voters.

In 2008 President-elect Barack Obama asked her to fill a new position as White House energy coordinator. Kathleen Rogers, president of the Earth Day Network, said, "Carol Browner's appointment as a 'climate czar' should signal to the world that the U.S. is taking a new direction on this issue." Browner is still fighting to protect our nation's water.

Pelageya Polubarinova-Kochina (1899-1999)

Pelageya Polubarinova-Kochina dedicated 75 years of her life to being a water scientist. She was born in Russia. Her parents wanted her to get a good education. So, when she was quite young, the family moved to St. Petersburg, the capital city of the Russian Empire.

Kochina graduated from high school in 1916. That fall she entered the University of Petrograd, today known as St. Petersburg State University. She stayed in school while her country was in the throes of World War I and the Russian Revolution. The revolution of 1917 overthrew the monarchy and replaced it with a communist government.

When her father died in 1918, Kochina got a job in the geophysical laboratory at the university. She studied and worked until tragedy struck again. She and her younger sister contracted tuberculosis,

27

a disease of the lungs. Her sister died, but Kochina recovered and went on to complete a degree in mathematics. After graduation, she stayed at the laboratory. While working there, she discovered a passion for fluid dynamics, the study of liquids and gases in motion. She also met Nikolai Kochin, her future husband. He worked for the Russian military. They married in 1925 and had two daughters.

Kochina left the laboratory to spend more time with her two young daughters. During those years, she taught at several schools and continued her research. In 1940 she earned a doctorate in physical and mathematical science.

After World War II ended, Kochina focused her research on using mathematical equations to understand the flow of liquids, especially water. She continued her research for the next 50 years. She wrote and lectured about the flow of groundwater and how it moves through porous materials. When she was 100 years old, she published her final paper on fluid dynamics and died two months later.

Did You Know?
Kochina co-founded the Siberian branch of the Academy of Sciences in Novosibirsk, Russia.

Environmental Engineer: Alexandria Boehm

Stanford University

Alexandria Boehm grew up in Hawaii, surrounded by the ocean. Behind her mother's house was a water canal where she swam and played every day. When the water in the canal became polluted, she decided to go to college and become an environmental engineer. She wanted to find ways to stop pollution and clean up dirty water.

Boehm moved to the mainland and went to college in California. After earning her doctorate, she was hired by Stanford University to teach and conduct research on ways to keep water clean. One way is by using tiny organisms, fungus, or green plants to attack and filter out pollutants. Doing this returns the water to its natural clean state. Today Boehm's work focuses on the health of coastal waters and new and innovative ways to keep it clean.

Boehm gets samples of ocean water. She takes them back to test them in her lab.

Boehm tests the ocean water for bacteria that can make swimmers sick.

Water Scientists at a Glance

Alexandria Boehm

Fields of study: *Civil and environmental engineering*

Known for: *Finding ways to prevent pollution and clean up already-polluted water*

Nationality: *American*

Birthplace: *Oahu, Hawaii*

Awards and honors: *Faculty Fellow Award from University of California, Irvine, 2000–2002; Pacific Rim Center for Oceans and Human Health Visiting Scholar, 2007; NSF Career Award, 2007; participant in National Academy of Engineers' Frontiers of Engineering Symposium, 2008*

Carol Browner

Field of study: *Environmental politics*

Known as: *Assistant to the president for energy and climate change*

Nationality: *American*

Birthplace: *Miami, Florida*

Date of birth: *December 16, 1955*

Awards and honors: *Longest-serving administrator of the Environmental Protection Agency, 1993–2001; chair of Audubon Society, 2003–2006*

Henry Darcy

Field of study: *Engineering*

Known for: *Building a pressurized water system that carried water from a spring through an aqueduct to reservoirs near Dijon and then to a network of pipes that delivered water into the city*

Nationality: *French*

Birthplace: *Dijon, France*

Date of birth: *June 10, 1803*

Date of death: *January 3, 1858*

Awards and honors: *Chief engineer for the département Côte-d'Or, 1858*

George Hadley

Fields of study: *Physics, meteorology*
Known for: *Proposing why trade winds occur*
Nationality: *English*
Birthplace: *London, England*
Date of birth: *February 12, 1685*
Date of death: *June 28, 1768*

Awards and honors: *Elected as a Royal Fellow, 1745; Hadley Centre for Climate Prediction and Research, crater on Mars named for him*

Edmond Halley

Fields of study: *Astronomy, meteorology*
Known for: *Calculating the orbit of Halley's Comet*
Nationality: *English*
Birthplace: *London, England*
Date of birth: *November 8, 1656*
Date of death: *January 14, 1742*

Awards and honors: *Lunar crater, Martian crater, Halley Research Station, street in Victoria, Australia, and comet named for him*

Mohammed Karaji

Fields of study: *Mathematics, hydrology*
Known for: *His studies on groundwater*
Nationality: *Iranian*
Birthplace: *Karaj, Iran*
Year of birth: *c. 953*
Year of death: *c. 1029*

Awards and Honors: *His school of algebra flourished for several hundred years after his death*

Bernard Palissy

Field of study: *Earth science*

Known for: *Discovering rain as the source of rivers and springs*

Nationality: *French*

Birthplace: *St. Avit, France*

Year of birth: *c. 1510*

Year of death: *c. 1589*

Awards and honors: *Named Inventor of Rustic Pottery to the King and Queen Mother; summoned by Catherine de Medici to decorate her palace, 1567*

Pelageya Polubarinova-Kochina

Field of study: *Mathematics*

Known for: *Her work in fluid dynamics*

Nationality: *Russian*

Birthplace: *Astrakhan, Russia*

Date of birth: *May 13, 1899*

Date of death: *July 3, 1999*

Awards and honors: *Stalin Prize, 1946; Hero of Socialist Labor, 1969; Order of the Friendship of Nations, 1979*

Glossary

air pressure—force exerted by the weight of the molecules that make up air; usually, the lower the air pressure, the stronger the storm

aquifers—layers of soil, rock, and other porous material through which water can flow

astronomy—study of the universe and of objects in space, such as the moon, sun, planets, and stars

atmosphere—blanket of gases that surrounds a planet

atom—smallest particle of an element

climate—conditions in the atmosphere in a particular place over periods of time

comet— icy and dusty object that orbits a star

elements—substances made of atoms with the same number of protons in their nuclei; cannot be broken down into simpler substances

engineering—work that uses scientific knowledge for practical things

equator—imaginary line drawn around the middle of Earth an equal distance from the North Pole and the South Pole

evaporation—process of change in water from a liquid to a gas

fossils—remains of ancient plants or animals that have hardened into rock; also the preserved tracks or outlines of ancient organisms

geophysics—use of physical principles to study the properties of Earth

groundwater—water found in underground chambers; it is tapped for drinking water through wells and springs

H_2O—chemical formula for water

hydrant—pipe that connects to a source of water

hydraulics—anything operated or moved by water or other liquid

hydrology—study of water (ice, water, water vapor), and how it circulates around Earth and in the atmosphere

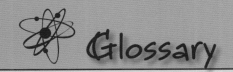
Glossary

molecules—small bits of matter made of two or more atoms bonded together

monsoons—weather seasons that are characterized by very heavy wind and rainfall

Protestants—members of Christian churches other than the Roman Catholic Church and Orthodox Church

qanaat—underground irrigation canal that is built when a farm or village needs water

reliefs—figures or shapes that are elevated from a flat surface

reservoirs—artificial lakes where water is collected

telescope—instrument made of lenses and mirrors that is used to view distant objects

trade winds—winds that blow in regular patterns

Water Science Through Time

c. 6000 B.C.	First irrigation systems and diversion dams are built in Mesopotamia and Egypt
c. 3050 B.C.	Water levels of the Nile River are first recorded in Egypt
c. 1800 B.C.	Natural lake in Egypt is used to store surplus water in the first reservoir
c. 1500 B.C.	Egyptians use the mineral alum to filter dirty water
c. 400 B.C.	First recorded mention of rainfall being measured in India
c. 150 A.D.	Roman inventor Heron of Alexandria creates a mechanical water fountain
c. 350	Greek philosopher Aristotle publishes *Meteorology*, where he hypothesizes that water vapor from soil condenses in mountain caverns, forming underground lakes and springs
c. 600s	Assyrians build the first aqueduct to move and store water
c. 8th century	Tunnels called qanaats are built and designed with the intent to transport water in ancient Turkey
1400	First rain gauges are used in Korea
1574	First flood warning system is developed in China

1580	Bernard Palissy writes that underground water comes from rainwater infiltrating the soil
1662	English scientist Sir Christopher Wren invents a mechanical rain gauge using a tipping bucket
1674	French hydrologist Pierre Perrault publishes *On the Origin of Springs*, where he refutes Aristotle's theory on underground water sources; instead, Perrault states that rainfall alone is enough to replenish the water flow in rivers
1687	Edmond Halley proves Perrault's theory that the amount of water evaporated from oceans and rivers creates enough precipitation to replenish the water flow of rivers
1814	On July 1, a deadly waterspout sinks a boat off Charleston, South Carolina, killing 40 crew members
1885	American Wilson Bentley uses a camera and a microscope to make the first photograph of a snowflake; later proposes that no two snowflakes are the same
1914	U.S. Public Health Service first begins regulating the quality of public drinking water; standards are expanded in 1925, 1946, and 1962

1960s	Scientists notice that snow cover around the world decreased by about 10 percent because of global warming
1969	Survey by the Public Health Service shows that only 60 percent of water systems deliver water that meets all the government standards
1980s	Scientists start trying to find a way to use the sun's energy to make hydrogen fuel out of water
2000	Scientists at NASA discover evidence of water on Mars
2009	California faces record droughts; thousands of farms are cut off from water in an attempt to preserve water sources

Additional Resources

Barnhill, Kelly Regan. *Do You Know Where Your Water Has Been? The Disgusting Story Behind What You're Drinking*. Mankato, Minn.: Capstone Press, 2009.

Flynn, Claire E. *Water World: Earth's Water Cycle*. New York: PowerKids Press, 2009.

Royston, Angela. *The Life and Times of a Drop of Water*. Chicago: Raintree, 2006.

Silverman, Buffy. *Saving Water: The Water Cycle*. Chicago: Heinemann Library, 2008.

Woodward, John. *Oceans*. New York: DK Publishing, 2008.

Internet Sites

FactHound offers a safe, fun way to find Internet sites related to this book. All of the sites on FactHound have been researched by our staff.

Here's all you do:
  Visit *www.facthound.com*
FactHound will fetch the best sites for you!

Index

aqueduct, 24
aquifer, 9, 10
atmosphere, 6, 14, 22

Boehm, Alexandria, 29
Browner, Carol, 26

canal, 10, 23, 29
comet, 6–7, 18, 19

Darcy, Henry, 23–25
diving bell, 20–21

evaporate, 14, 21, 22

fluid dynamics, 28

geophysics, 20, 27
groundwater, 9, 11, 25, 28

Hadley, George, 22
Halley, Edmond, 18–21, 22
hydrology, 9, 25

Karaji, Mohammed, 8–9

molecule, 6, 7

Palissy, Bernard, 12–17, 21
Polubarinova-Kochina, Pelageya, 27–28

qanaat, 10

reservoir, 24, 25
river, 12, 14, 16, 20, 21, 23

streams, 14, 16, 23

trade winds, 22
tunnel, 10, 11

water
  gas, 7, 28
  liquid, 6, 7, 28, 29
  solid, 7
water cycle, 12, 14
weather, 17, 18, 20
  monsoon, 20
  rain, 12, 14, 15, 18, 21
  snow, 14

About the Author

William B. Rice

William Rice grew up in Pomona, California, and graduated from Idaho State University with a bachelor of science in geology. For 18 years, he has worked at a California state agency that works to protect the quality of surface and groundwater resources. He has overseen the evaluation and cleanup of pollution in groundwater and water in rivers, lakes, and streams. Protecting and preserving the environment is important to him. He is married with two children and resides in Southern California.

Image Credits

North Wind Picture Archives/Alamy, cover (left), 18, 31; Terry Ashe/Time Life Pictures/Getty Images, cover (right), 26 (inset), 30 (middle); Tomasz Szymanski/Shutterstock, 3, 26; Photos.com, 6; Tim Bradley, 7, 9 (bottom), 14, 22, 24; The Granger Collection, New York, 8, 17; The University of Texas at Austin, 9 (top); Jesper Jensen/Alamy, 10; Brian McEntire/Shutterstock, 11 (top); Julius Costache/Dreamstime, 11 (bottom); Mary Evans Picture Library/Alamy, 12, 21, 32 (top); Corbis, 13; Andrew F. Kazmierski/Shutterstock, 19; NASA, 20; Tibor Bognar/Alamy, 23 (right); Courtesy of Henry Darcy Family, 23 (inset), 30 (bottom); Courtesy of Glenn Brown, 25 (all); Zastavkin/Shutterstock, 27 (left); Rick Reason, 27 (inset), 32 (bottom); Brian Tan/Shutterstock, 28; Phil Degginger/Getty Images, 29 (left); USDA/Susan Boyer, 29 (bottom right); Stanford Univesrity, 29 (top), 30 (top).